First Facts™

Exploring the Animal Kingdom

Amphibians

dainty green tree frog

by Adele Richardson

Consultant:
Robert T. Mason
Professor of Zoology, J. C. Braly Curator of Vertebrates
Oregon State University
Corvallis, Oregon

Capstone press

Mankato, Minnesota

W9-DEX-696

First Facts is published by Capstone Press,
151 Good Counsel Drive, P.O. Box 669, Mankato, Minnesota 56002.
www.capstonepress.com

Library of Congress Cataloging-in-Publication Data
Richardson, Adele, 1966–
    Amphibians / by Adele Richardson.
    p. cm.—(First facts. Exploring the animal kingdom)
    Includes bibliographical references and index.
    ISBN-13: 978-0-7368-2620-4 (hardcover)
    ISBN-10: 0-7368-2620-3 (hardcover)
    ISBN-13: 978-0-7368-4941-8 (softcover pbk.)
    ISBN-10: 0-7368-4941-6 (softcover pbk.)
    1. Amphibians—Juvenile literature. I. Title. II. Series.
QL644.2R5275 2005
597.8—dc22   2004000608

Summary: Discusses the characteristics, eating habits, and offspring of amphibians, one of the
    main groups in the animal kingdom.

Editorial credits
Erika L. Shores, editor; Linda Clavel, designer; Kelly Garvin, photo researcher; Eric Kudalis,
    product planning editor

Photo credits
Bruce Coleman Inc./E. R. Degginger, 18–19; Michael Fogden, 6–7, 11 (top left), 15
Corbis, cover (main right); B. Borell Casals/Frank Lane Picture Agency, 20
Corel, cover (top left, bottom left), 11 (bottom)
DigitalVision/Gerry Ellis and Michael Durham, 12–13; PictureQuest, 1
Dwight R. Kuhn, 9, 16
James P. Rowan, 11 (top right)
Minden Pictures/Mitsuhiko Imamori, 17
Photodisc Inc., cover (middle left)

Table of Contents

Amphibians

Amphibians are a group in the animal kingdom. **Caecilians**, toads, and salamanders are amphibians.

Other groups of animals live on earth with amphibians. Birds have feathers. Reptiles have thick, dry skin. Fish breathe with **gills**. Insects have six legs. Mammals have hair.

Birds

Mammals

Reptiles

Main Animal Groups

Insects

Amphibians

Fish

Amphibians Are Vertebrates

Amphibians are **vertebrates**. Vertebrate animals have backbones. A backbone is made up of smaller bones called **vertebrae**. Caecilians have many vertebrae. Caecilians can bend easily.

Mexican burrowing caecilian

Amphibians Are Cold-Blooded

Amphibians are cold-blooded. Their body temperatures change with the temperature around them. Amphibians must not get too hot or too cold. They sit under plants to cool off. This salamander lies in the sun to warm up.

Fun Fact!
Many amphibians rest during the day. They are more active at night when it is cooler.

spotted salamander

Amphibian Bodies

Amphibians come in many shapes. Caecilians have small eyes and no legs. Rings along their bodies make them look like worms. Frogs have short bodies and no tails. Salamanders have long bodies and tails. Both frogs and salamanders have four legs.

varagua caecilian

wood frog

marbled salamander

blue poison dart frog

Moist Skin

Most amphibians have smooth, moist skin. **Mucus** covers their skin. Mucus keeps the skin of this frog and other amphibians wet and slippery.

Amphibians outgrow their skin. Old skin peels off when it becomes too small. New skin grows in its place.

Fun Fact!

Some amphibians make a poison that covers their skin. The poison dart frog's poison is strong enough to kill its enemies.

How Amphibians Breathe

Most amphibians breathe air in two ways. Most adult amphibians breathe with **lungs**. People breathe air in this way. Amphibians also breathe through their skin. **Oxygen** can go through this tree frog's skin and into its blood.

Reinwardt's tree frog

spotted salamander

What Amphibians Eat

Amphibians catch and eat other animals for food. Amphibians eat mostly worms, insects, and snails.

Japanese toad

Most amphibians do not move very fast. They sit and wait for animals to come to them. This toad catches an insect with its long, sticky tongue.

Young Amphibians

Most amphibians lay eggs in water. Frogs and toads **hatch** from their eggs as **tadpoles**. Tadpoles breathe underwater with gills. Tadpoles' bodies go through many changes before they become adults.

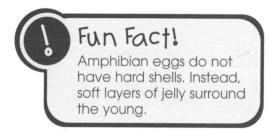

Fun Fact!
Amphibian eggs do not have hard shells. Instead, soft layers of jelly surround the young.

Amazing but True!

The male midwife toad is a caring father. The female toad lays two strings of eggs. The male toad then wraps the eggs around its back legs. The male toad carries the eggs until they hatch. It even dips the eggs in water at night to keep them moist.

Compare the Main Animal Groups

	Vertebrates	Invertebrates	Warm-blooded	Cold-blooded	Hair	Feathers	Scales
Amphibians	X			X			
Birds	X		X			X	
Fish	X			X			X
Insects		X		X			
Mammals	X		X		X		
Reptiles	X			X			X

Glossary

caecilian (si-SIL-yuhn)—a wormlike amphibian with no legs

gill (GIL)—a body part on the side of a tadpole or fish that helps it get oxygen from water

hatch (HACH)—to break out of an egg

lungs (LUHNGS)—organs inside the chest that animals use to breathe

mucus (MYOO-kuhss)—sticky, wet liquid made by glands to protect parts of the body

oxygen (OK-suh-juhn)—a colorless gas in the air and water that animals need to breathe

tadpole (TAD-pohl)—the stage of a frog's or toad's growth between the egg and adult stages; tadpoles live in water.

vertebrae (VUR-tuh-bray)—small bones that make up a backbone

vertebrate (VUR-tuh-bruht)—an animal that has a backbone

Read More

Harvey, Bev. *Amphibians.* Animal Kingdom. Philadelphia: Chelsea Clubhouse Books, 2003.

Mattern, Joanne. *Reptiles and Amphibians.* The Rosen Publishing Group's Reading Room Collection. New York: Rosen, 2003.

Parker, Edward. *Reptiles and Amphibians.* Rain Forest. Austin, Texas: Raintree Steck-Vaughn, 2003.

Internet Sites

FactHound offers a safe, fun way to find Internet sites related to this book. All of the sites on FactHound have been researched by our staff.

Here's how:
1. Visit *www.facthound.com*
2. Type in this special code **0736826203** for age-appropriate sites. Or enter a search word related to this book for a more general search.
3. Click on the **Fetch It** button.

FactHound will fetch the best sites for you!

Index